To our readers, those students just beginning a life of adventure and learning, always remember:

You are braver than you believe, stronger than you seem,
and smarter than you think.
~
A.A. Milne

More Sounds of the Night:
A Child's Interactive Book of Fun & Learning

Written & Created by

Brent A. Ford

ISBN 978-1-947348-76-9

nVizn
ideas

www.nviznideas.com

Interactive Components

This version of **Sounds of the Night** combines the best of both worlds. It is a physical book where children can turn the pages, gaze at the photographs and sit close to a parent or loved-one. It is also a book featuring tech-based, interactive components to extend the fun and the learning.

To access the web-based features, use a mobile device (phone or tablet) with nVizn's QR Code Reader. Our app is FREE and does NOT include advertising or in-app purchases. Our system is also designed to be "kid friendly" - meaning that the app does not open up the entire Internet to young children. Our app only reads codes created by us, so users can access only web content that we create and maintain. Look for the nVizn QR Code Reader in your app store.

After downloading the code reader, simply open the app on your phone or tablet, point it at any one of the many codes throughout the book and you are off. The code reader will automatically take you to a webpage for some learning and fun! (You will need an Internet connection to access these features.)

nVizn Ideas

QR Code Reader

Links to Learning

← Swipe left to begin

Try it now....open your QR Code reader and point it at this code.

Remember a time outside when dusk turned to night.

Remember climbing that tree
to see the sunset.

Remember the sounds of the night.

Think back. What did you hear?

Listen now.

Ever wonder about those sounds?

How many different sounds do you hear?

Listen again.

Different animals make those sounds.

Let's learn.

What animal is making this sound?

It is a longwinged katydid calling out to other katydids.

Can you hear the katydids now within all the nighttime sounds?

What animal is making this sound?

It is a chorus of Pacific treefrogs singing to one another.

Can you hear the Pacific treefrogs now within all the nighttime sounds?

What animal is making this sound?

It is a spotted bat flying around looking for food.

Can you hear the spotted bat now
within all the nighttime sounds?

What animal is making this sound?

It is a wolf howling — probably not at the moon, but to other wolves nearby.

Can you hear the wolf now within all the nighttime sounds?

What animal is making this sound?

It is a barn owl calling out
to other owls in the night.

Can you hear the barn owl now
within all the nighttime sounds?

What animal is making this sound?

It is a field cricket calling out to other crickets. Can you hear the crickets now within all the nighttime sounds?

What animal is making this sound?

It is a coyote calling out to its family.
Can you hear the coyote now within all
the nighttime sounds?

What animal is making this sound?

It is a bird called a nighthawk. It does not look much like a hawk, does it?

Can you hear the nighthawk now within all the nighttime sounds?

What animal is making this sound?

It is a little pine squirrel.

Can you hear the pine squirrel now
within all the nighttime sounds?

What animal is making this sound?

It is a night heron calling out to other herons.

Can you hear the night heron now within all the nighttime sounds?

What animal is making this sound?

That is a bobcat. It sounds like that bobcat is really mad about something, doesn't it?

Can you hear the bobcats now within all the nighttime sounds?

What animal is making this sound?

It is an elk. If you listen closely
you will hear other elk calling back.

Can you hear the elk now
within all the nighttime sounds?

Equipped with five senses, we explore the universe ...

.... around us and call the adventure science.

Edwin Powell Hubble

Now when you listen to the sounds of the night, what do you hear?

Learn with Simon

Hi, my name is Simon.

Never seen an animal like me? I am an indri (pronounced IN dree) and I live in a place called Madagascar. To learn more about indris and Madagascar, follow this QR code.

I'm nVizn's mascot - the nVizn indri - and I will be your guide as we learn about the world in which we all live. We'll watch a video or two, do an activity or two, and learn to think and work like a scientist. I'll help get you started on a life-long process of learning about how our world works AND why it works as it does.

So, let's get started!

Sounds of the Night

From chirping katydids to croaking frogs and singing birds, our world is filled with the sounds of nature. The variation in the sounds animals make, and how they make those sounds, is truly awe-inspiring.

While there are many different kinds of sounds, we should remember (or learn) that sound is created when something vibrates (moves back and forth really fast). When we yell or sing or hum, there is a part of our body (a part inside our throat) that we make vibrate. How do we know that? Put your hand on your throat....blow out without making a noise. Now put your hand on your throat and sing, hum, yell. Is there a difference? Sure there is...you do not feel anything when you just blow out and you feel something vibrating in your throat when you make a sound. You are causing that vibration, which we hear as singing or yelling or humming. If we can make sound by causing our throats to vibrate, does it make you wonder how other animals make sounds? Follow this QR code to see three different animals - a cricket, a toad and a cicada - and learn how they make sound.

Bats Aren't Blind

There's a saying we often hear - *blind as a bat* - which perhaps originated from the fact that bats are active mainly at night. Who needs to see if it's dark? In fact, bats are not blind and can see quite well. However, bats have developed a way to fly around in the dark and it does not involve superior eyesight like that of an owl. Bats use sound to help them fly at night in search of food. Follow this QR code to learn more about how bats use sound.

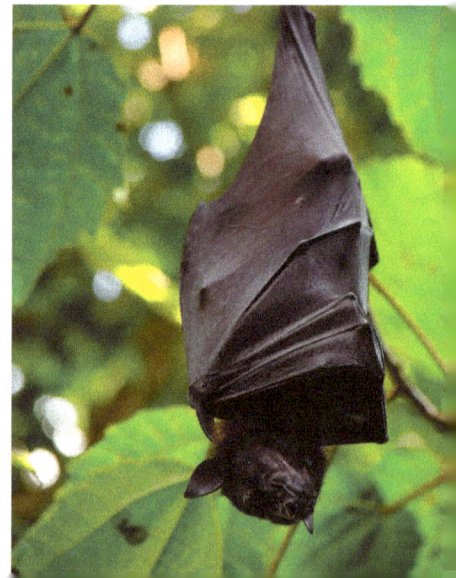

Animals & Music

In this book, we learned a lot about animals and the sounds that they make. But Simon wonders if you have ever thought about how animals respond to other sounds - like music. We know how humans respond to music - they love it! For thousands of years, humans have created music ... from simple drum beats to grand orchestral music ...from rock-and-roll to hip-hop songs. Follow this QR code to see evidence that animals love music too.

Coyote Communication

Of all the animals on Earth, coyotes are among the most expressive in the range of sounds they make. While scientists have ideas about what certain coyote sounds mean, there is still a lot of study to be done. Follow this QR code to learn more about how coyotes use sound to communicate with others. Maybe you will be the scientist that decodes coyote language.

nVizn Ideas

Did You Use Soap?
Written by Brent A. Ford
Illustrations by Kristina Muñoz
Interactive

A World of Wonder
A Child's Interactive Book of Wonder
Brent A. Ford
Lucy McCullough Hazlehurst
Interactive

When We Were Young
A Story of Dr. Doolittle's Pushme-Pullyou
Lucy McCullough Hazlehurst & Brent A. Ford
Kristina Muñoz

When the Cows Come Home
Written by Brent A. Ford & Lucy McCullough Hazlehurst
Illustrations by Seokwon Kim

Sounds of the Night
A Child's Interactive Book of Fun & Learning
Brent A. Ford
Interactive

The Water Thief
Brent A. Ford & Lucy McCullough Hazlehurst
Interactive

A Bluebird Weather Day
Written by Brent A. Ford & Lucy McCullough Hazlehurst
Illustrations by Seokwon Kim

Let the Dark Out
Written by Brent A. Ford & Lucy McCullough Hazlehurst
Illustrations by Seokwon Kim

Well, Will You Look at That?

Marcus & the Kudzu Monster
Lucy McCullough Hazlehurst & Brent A. Ford
Seokwon Kim

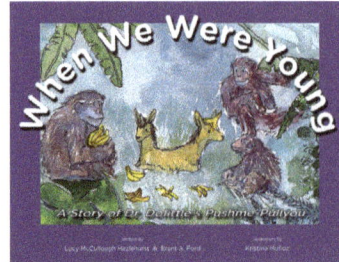

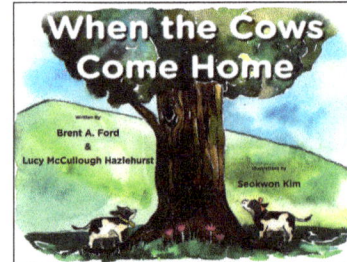

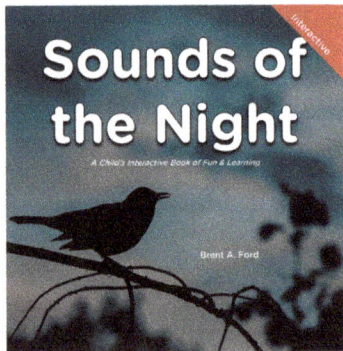

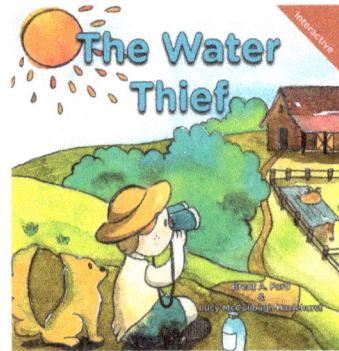

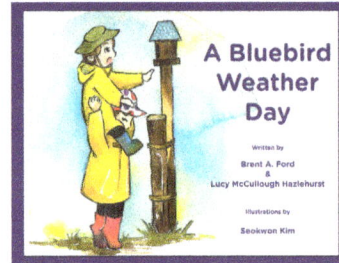

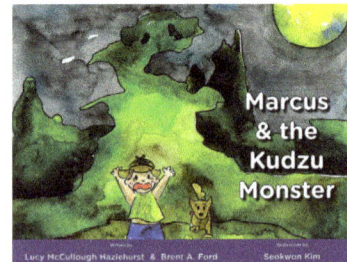

To learn more about these and other children's books that include tech-based resources, follow this QR code.

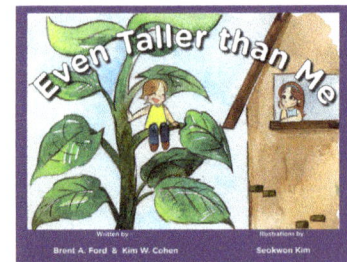

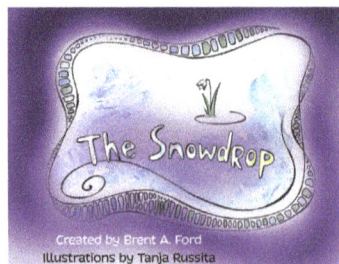

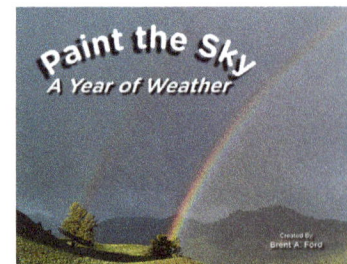

When I Grow Up

Even Taller than Me
Written by Brent A. Ford & Kim W. Cohen
Illustrations by Seokwon Kim

The Snowdrop
Created by Brent A. Ford
Illustrations by Tanja Russita

Paint the Sky
A Year of Weather
Created By Brent A. Ford

www.ingramcontent.com/pod-product-compliance
Lightning Source LLC
Chambersburg PA
CBHW051322020426
42333CB00031B/3443